UNDERSTANDING NATURAL DISASTERS

DROUGHTS

by Sue Bradford Edwards

BrightPoint Press

San Diego, CA

© 2025 BrightPoint Press
an imprint of ReferencePoint Press, Inc.
Printed in the United States

For more information, contact:
BrightPoint Press
PO Box 27779
San Diego, CA 92198
www.BrightPointPress.com

LIBRARY OF CONGRESS CATALOGING-IN-PUBLICATION DATA

Names: Edwards, Sue Bradford, author.
Title: Droughts / by Sue Bradford Edwards.
Description: San Diego, CA: BrightPoint Press, [2025] | Series: Understanding natural disasters | Includes bibliographical references and index. | Audience: Grades 7-9
Identifiers: LCCN 2024013380 (print) | LCCN 2024013381 (eBook) | ISBN 9781678208684 (hardcover) | ISBN 9781678208691 (eBook)
Subjects: LCSH: Droughts--Juvenile literature. | Drought management--Juvenile literature. | Plants--Effect of drought on--Juvenile literature.
Classification: LCC QC929.25.E393 2025 (print) | LCC QC929.25 (eBook) | DDC 363.34/929--dc23/eng/20240423
LC record available at https://lccn.loc.gov/2024013380
LC eBook record available at https://lccn.loc.gov/2024013381

CONTENTS

AT A GLANCE

- A drought is a period when an area is drier than normal.

- Droughts can last for weeks or even years.

- Droughts happen when weather patterns change. Changes in weather patterns can occur naturally. Human actions can also contribute to these changes.

- In 2018, the United Nations (UN) reported that 40 percent of the world's population was impacted by water shortages.

- When droughts happen, farmers struggle to grow crops. This can lead to food shortages.

- Droughts can lead to wildfires and dust storms.

- During a drought, governments may restrict how much water people can use. This ensures that everyone can get the water they need.

THE DUST STORM

It was the morning of May 1, 2023. The day was sunny. The wind gusted up to 45 miles per hour (74 kmh). Cars were driving on Interstate 55 in Illinois. The highway stretched between farm fields. The area was flat with very few trees.

A huge cloud of dust approached the highway. Witnesses later described the cloud as being 200 feet (61 m) high.

The National Weather Service advises drivers caught in a dust storm to pull off the road and turn off taillights to avoid confusing other drivers.

One car crash at high speeds can cause several more. This is known as a multi-vehicle accident or a pile-up.

Soon the highway was enveloped in a dust storm. Drivers couldn't see ahead of them.

Evan Anderson was one driver caught in the storm. He was driving home to St. Louis, Missouri, from Chicago, Illinois. He said, "It was like a white out, only it was a brown out." A white out happens when swirling snow makes it impossible for people to see. The dust in the air had the same effect. Anderson explained his situation to reporters. "People tried to slow down, and other people didn't, and I just got plowed into," he said. "There were just so many cars and semitrucks with so much momentum behind them."[1]

Anderson's car was trapped between two semitrucks. The driver of the rear truck turned slightly. The truck hit Anderson. But it also blocked other cars from hitting him. Eighty-four cars were involved in the accident.

The dust storm resulted from weeks without rain. A drought had dried out the soil. When farmers plowed their fields, they

Crops affected by drought are more likely to also suffer from pests and disease.

loosened the soil. Strong winds kicked up the soil and created the dust storm.

THE DANGER OF DROUGHTS

Droughts occur when the land dries out. This happens because of a lack of rain and snow. Human activity can also make droughts worse. People can use too much water. Sometimes they do things that dry out the land.

During a drought, people, plants, and animals can't get the water they need. Farmers might not be able to give their crops enough water. This can cause food shortages. However, people can defend against droughts. They can **conserve** water before and during droughts. This keeps water available for when it is needed.

CHAPTER ONE

WHAT ARE DROUGHTS?

Droughts happen when an area doesn't get enough **precipitation**. When a drought occurs, the soil dries out. In extreme droughts, plants and animals die. The soil becomes cracked. People may not have access to the water they need. This makes it hard to grow food. Droughts can also lead to fires and dust storms.

A drought in one area may not be considered a drought in another area.

Droughts can lower the water level of bodies of water, exposing layers of rock typically not seen.

The driest areas in the United States, such as Death Valley National Park in California and Nevada, usually receive only a few inches of rain per year.

It depends on the area's normal amount of precipitation. Tampa, Florida, regularly gets about 46.3 inches (117.6 cm) of rain per year. If less rain falls for a long time, a drought could occur.

Other places usually get less rain than Tampa. Albuquerque, New Mexico, averages about 9.5 inches (24 cm) of rain and snow per year. If Tampa got that much precipitation, it would be at risk for drought. But this would be normal for Albuquerque.

Some droughts last for only a few weeks. Others last for months. The worst last for years. It is hard to say exactly when a

Plants Adapt

Periods of drought and dry conditions are normal. Plants have adapted ways to survive droughts. Water evaporates from a tree's leaves. Some trees, including the Southern Magnolia, shed leaves. Dropping leaves reduces evaporation. Some types of grass, such as Kentucky bluegrass, turn brown and go **dormant**.

drought begins. A week without rain isn't a drought. But several weeks without rain can be a drought. In the United States, experts decide when a water shortage becomes a drought.

Lack of rain can lead to soil conditions that cause flooding. "We expect that . . . the grasses and trees would die during the drought," says meteorologist Mark Elliot. A meteorologist is a scientist who studies the weather. Elliot works for The Weather Channel. "But it's not just about what happens above the soil. We're talking about the soil itself," he says.[2]

Soil compacts during a drought. Water is a normal part of soil. It fills gaps between **particles** of soil. During a drought, the water in the soil evaporates.

SOIL COMPACTION

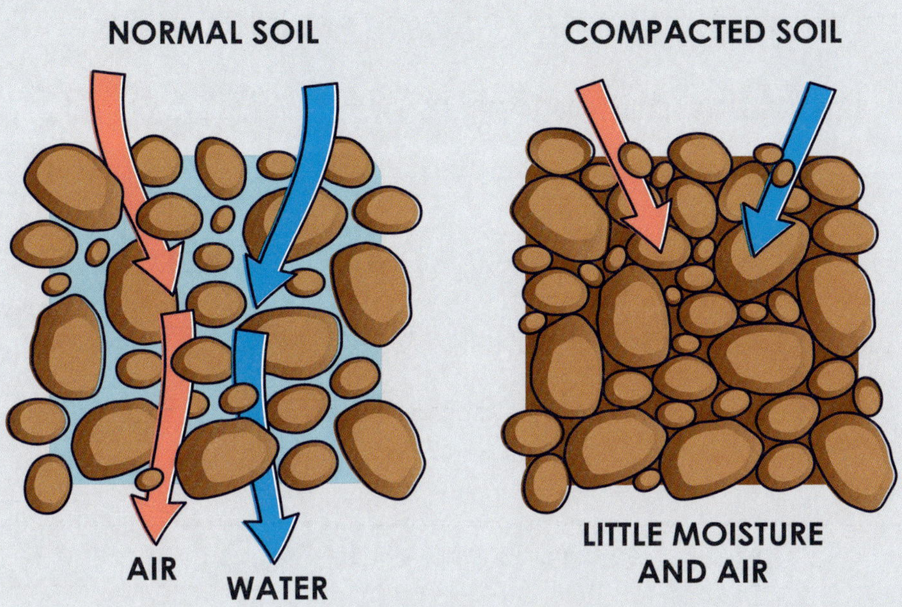

Compacted soil does not let air and water through the way normal soil does. It can be hard for plants to grow in compacted soil.

The soil dries out. Then it shifts. The soil particles move closer together. This is called compaction. Soil can shift for many reasons. Animals walk on it. Trucks and farming equipment drive over it. Even the impact of rain can compact soil.

Compacted soil shrinks. It pulls away from the foundations of buildings. The surface of the soil cracks. Compacted soil is hard like concrete. It cannot absorb rain. Instead, the rain pools. It gathers and then runs off. This can cause floods.

WHAT CAUSES DROUGHTS?

Droughts occur when weather patterns change. This can happen naturally. A warm, dry summer can cause a drought. People can change weather patterns, too. This is one consequence of climate change, which is primarily caused by human activity. People release pollution into the **atmosphere**. This pollution traps more of the sun's heat in the atmosphere than usual. Less heat escapes into space. This raises global temperatures. Warmer temperatures change weather patterns.

Deforestation is the removal of trees from an area. Sometimes people cut down forests to build or farm on the land. This can cause droughts. Trees hold water.

One reason people cut down forests is to make space for crops such as palm trees, from which palm oil can be harvested.

They release it into the atmosphere.
Removing trees interrupts this cycle.

People also cause drought by using too much water. This might happen because more people move into an area. Or it may happen after farming in an area increases. Building dams can also lead to drought. Dams interrupt the natural flow of water.

THE WATER CYCLE

Changes in weather patterns alter the water cycle. The water cycle is the ongoing movement of water on Earth. Liquid water evaporates and becomes water vapor. Under certain conditions, the vapor condenses. Condensation is the opposite of evaporation. Water droplets form clouds. When this water falls back to Earth's

Under the right conditions, droplets of water gather in storm clouds. When these droplets grow too large, they fall to the earth as rain.

surface, it is called precipitation. Rain and snow are familiar forms of precipitation.

Changing weather patterns can prevent storms that would bring rain. El Niño is a weather pattern. It occurs when water on the surface of the Pacific Ocean heats up.

During the Dust Bowl, dust storms swept across the
Great Plains region of the United States.

The warmer water changes storm patterns, resulting in reduced rainfall in Indonesia, Australia, and northeastern South America. These areas become at risk for droughts. El Niño occurs every 2 to 7 years. Climate scientists cannot always tell when it will happen next.

La Niña is another weather event. It takes place when the same Pacific waters cool off. La Niña also alters storm patterns. It causes reduced rainfall for a different set of regions. This can lead to drought in North and South America. La Niña and El Niño each last about a year. Scientists believe that La Niña caused the Dust Bowl. This was a historic North American drought that lasted from 1930 to 1936.

WHERE DROUGHTS HAPPEN

Droughts can happen anywhere in the world. António Guterres explained in a 2022 speech that droughts across the world were becoming more common and intense. Guterres is the secretary-general of the United Nations (UN).

The UN is an organization of countries from around the world. They work together to solve worldwide problems. The UN reported in 2018 that 40 percent of the world's population was affected by water shortages. Some countries are hit harder by droughts than others. In 2022, these countries included Afghanistan, Angola, Brazil, Burkina Faso, Chile, Ethiopia, and Iraq. Guterres explained that countries that experience droughts have sand and

Wildfires become more common during droughts because dry vegetation acts as fuel for fires.

dust storms. He says that drought-affected areas have more wildfires and crop failures. People are often forced to leave their homes.

Droughts are a normal part of global weather patterns. Because droughts have a widespread and lasting impact, drought conditions are closely watched.

Scientists measure the moisture present in soil to monitor drought conditions.

A weekly map of drought conditions in
the United States is posted on the US
Drought Monitor. The monitor is the work of
many experts.

The US Drought Monitor does not create
a weather forecast. Weather forecasts
predict upcoming weather such as rain.
Instead, the US Drought Monitor map looks
back at recent weather. The map makes
it easy to see which areas are having low
rainfall. This helps scientists determine
when a drought starts and when it ends.
Scientists still have a lot to learn about
droughts. But the effects of drought
are clear.

THE EFFECTS OF DROUGHTS

Droughts can sometimes last for months. These events impact the individuals who live in affected areas. They also affect communities as a whole. Finally, droughts can have serious effects on the environment.

The World Health Organization (WHO) estimates that droughts affect 55 million people every year. During a drought, people

Even saguaro cacti, which are used to dry conditions, can die due to drought.

have water **insecurity**. This means they have a hard time finding water.

HOW DROUGHTS AFFECT INDIVIDUALS

Water insecurity is an especially serious problem for farmers. The Biggs family lived on a farm in West Goshen, California. Droughts in the state had dried out their well. By 2022, the well had been dry for more than 10 years. A well is a hole drilled in the ground. Wells reach underground sources of water. At first, the Biggs family saved water in underground tanks called cisterns. As the drought wore on, they had to drive into town to do their laundry. They visited neighbors who were willing to share water. Then they returned home

The underground layers from which wells draw water are called aquifers.

with full water containers for showering and drinking. They could not grow crops or raise animals.

Large farms and factories can make water shortages worse. They pump water out of very deep wells. The water they take is not available for individual farmers or towns. If people don't have access to water where they live, they leave. They must find new places to live.

HOW DROUGHTS AFFECT COMMUNITIES

Droughts also affect entire communities. When farmers struggle to grow crops, communities may have a hard time finding enough food. The food grown in their area may not be very good. It may also cost more money.

Leaves of wilting plants turn brown because they stop producing chlorophyll, a green chemical that helps plants grow in sunlight.

Jim Abma is a farmer. He explains how drought affects his crops. "That comes to smaller ears of corn, smaller tomatoes. Just the result of a stressed-out plant," says Abma.[3] Without enough water, the fruits and vegetables are not as good as what he usually grows.

Farmers pay for the water they use. During droughts, the cost of water increases. Even if the food crops are poor quality, they cost more to grow. People must pay more for this food.

Droughts can also affect a community's access to electricity. Running water produces 7 percent of the electricity used in the United States. To make this electricity, dams store water. The water is released to flow through turbines. Movement within these turbines is turned into electricity.

The Hoover Dam is on the Colorado River. It is between Nevada and Arizona. It is one of the largest facilities that uses water to create electricity. But there are also many smaller power plants that use water to make electricity.

When there is a drought, less water is available to generate electricity in this way. The price of energy rises. This happened in central California in March 2022. The Pacific Gas and Electric Company encouraged people to use less water. That way the water could be used to make electricity for their homes.

When water levels drop too far, dams cannot generate electricity. Old power

The Hoover Dam was completed in 1936. At the time, it was the world's tallest dam.

plants must be restarted. These plants use oil, coal, and natural gas to create electricity. Using these fuels produces **greenhouse gases**. Greenhouse gases worsen climate change.

When there is a drought, plants dry out and die. Dried plants provide fuel

Gas-powered cars release greenhouse gases by burning gasoline.

for wildfires. Some fires start with a lightning strike. Others start after electrical equipment fails. Even a campfire can quickly get out of control. Once a wildfire begins, it is hard to put out. Part of the problem is that fires can create winds that make them spread faster. "If it isn't that windy, but your fire is intense and releases a lot of heat, it has the potential to generate its own winds," says scientist Adam Kochanski. "This fire will start moving as if it were really windy."[4] Kochanski studies the atmosphere at the University of Utah.

Wildfires burn forests and other wild areas. But these fires can spread to populated areas. In November 2018, the Camp Fire began burning in northern California. It was the most destructive fire in California history. It occurred during a drought. The fire burned about 240 square miles (622 sq km). It destroyed almost 14,000 buildings and killed 88 people. Survivors had to find new places to live.

HOW DROUGHTS AFFECT THE ENVIRONMENT

Droughts also affect the environment. In 2022, Utah suffered from drought. It led to

Cultural Impacts of Wildfires

People living in remote areas can suffer from wildfires caused by droughts. The Karuk people live in northern California. They are American Indians. Fires in 2020 damaged lands where they hunt and fish. They could not visit certain areas. They could not hold traditional ceremonies.

The Great Salt Lake is too salty for fish, but certain microorganisms thrive in the salty water.

a 1-foot (0.3 m) drop in the water level of the

Great Salt Lake. This drought affected

the plants and animals that live in the lake.

Bonnie Baxter is a biologist. She studied

the impact of drought on **microorganisms**

that live in the Great Salt Lake. They live

on rocks that are normally underwater.

During the drought, these microorganisms

struggled. "You can see that they're dry

and they're not green and they're out of

Ponderosa pines grow in the western United States. One of their common pests is the mountain pine beetle.

the water. Even the ones in the water are not healthy, because they're too salty. The ones out of the water are too dry," said Baxter.[5] As the lake level dropped, the salts and minerals in the water became more concentrated. This was bad for the microorganisms that depended on a very specific balance.

Droughts can also lead to insect **infestations**. Healthy plants fight off insects. For example, a healthy ponderosa pine can stop an invading beetle. When the beetle chews a hole in the tree's bark, sap runs into the hole. It pushes the beetle back out. But drought stresses plants. The plants become less healthy. They cannot fight off insects. A stressed ponderosa pine cannot push beetles out. The beetles invade and kill the tree.

During a drought, plants die. This exposes and loosens soil. Farming can make this worse. High winds can pick up loose soil and cause a dust storm. That's what happened along Interstate 55 in Illinois in May 2023. When a dust storm engulfs a highway, accidents happen.

DEFENDING AGAINST DROUGHTS

The US Drought Monitor defines five levels of drought. The lowest level is Abnormally Dry. In these conditions, the growth of crops and pastures slows down. Pastures are the fields where livestock graze. There is also a raised fire risk.

The next level is Moderate Drought. In this stage, there is damage to crops and pastures. There is a high fire risk. Water sources are low. People may be asked to

Sometimes governments enforce water rationing during droughts. Rationing is control over the use of something scarce.

THIS AREA
HAS LIMITED
WATER
SAVE WATER
THIS SUMMER
We can't rely on rains to come.
Every drop counts.
Western Cape
Government
BETTER TOGETHER.
For water saving tips visit
www.h2Ohero.co.za

use less water. In a Severe Drought, crops will probably die. Fire risk is even higher. Water becomes scarcer. Governments limit how much water people can use.

In the Extreme and Exceptional Drought levels, conditions get even worse. At the Exceptional Drought stage, wells run dry. People look for new ways to save water. An Exceptional Drought struck Sedan, Kansas, in September 2023. Schools there had to shut off drinking fountains. People looked for new sources of water. The local movie theater saved the water that condensed in its air conditioner. The theater let people come and take some of the water. Fortunately, there are usually things that people can do before droughts get this serious.

Water condenses on air conditioners because the parts inside that cool the air absorb moisture from the air.

HOW PEOPLE CAN HELP

To reduce the impact of droughts, people need to act before a drought occurs. The most important step is conserving water. This means that, indoors and outdoors, people should not waste water.

A big part of indoor conservation is limiting how much water runs down the drain. People should capture the water they don't use. For example, people can place a container in the shower to capture the water as they wait for it to warm up. They can use this water for other purposes, such as watering plants. In drought-prone areas,

A faucet that leaks a drop of water every second can waste more than 3,000 gallons (11,000 L) in a year.

experts advise that people should not let showers run. People should get wet and then turn off the shower. After they lather up, they can turn the water back on to rinse. Experts also advise people to fix leaks. This means that dripping faucets need to be repaired or replaced. A toilet tank that leaks needs to be fixed.

People should also conserve water outdoors. This means not washing the car at home. Commercial car washes are more efficient, recycling their water. People should not wash grass clippings off the driveway. Instead, they should use a broom. People should also avoid overwatering their lawns and gardens. Finally, pool owners should cover their pools when not in use. This keeps the water from evaporating.

Conservation is especially important in the summer. Because summer weather is hot, people use more water. They use even more during droughts. John Nielsen-Gammon is a professor. He studies climate at Texas A&M University. In 2023, large areas of Texas experienced drought. Nielsen-Gammon said, "Continued high temperatures mean that plants lost water more rapidly than normal."[6] This is because heat causes the water to evaporate. He explained that people will then use even more water on gardens and crops.

DURING A DROUGHT

During a drought, people need to work even harder to conserve water. The United States

The average American family uses more than 300 gallons (1,100 L) of water per day.

Geological Survey offers a variety of tips. The first piece of advice is simple. People should always obey local water restrictions. These restrictions are designed to conserve water for essential uses. Water is necessary for life. People and animals need to drink water. It is needed to fight fires. People need water to cook food and wash clothing. Bathing is also necessary.

An average lawn sprinkler uses about 16 gallons (61 L) of water per minute.

During a drought, non-essential water use may be limited. This might mean restrictions on watering lawns. People may be asked not to flush toilets to throw away tissues. Or they may be asked to run dishwashers only when they are full.

Dallas, Texas, passed a water conservation law in 2012. Under this law, lawn watering is limited during droughts. The days people can water their lawns are determined by their house number. No one can water on Sunday or Thursday. Sprinklers that allow water evaporation cannot be used between 10 a.m. and 6 p.m. from April 1 to October 31. The city charges higher rates for higher water usage. This encourages people to use less water.

In 2014, cities in California also imposed restrictions. Kelley Dyer was a water supply analyst for Santa Barbara, California. She explained some of the restrictions to the public. She said, "All hoses are required to have a shut-off nozzle. . . . There [are] restrictions on operation of ornamental fountains. All pools and spas are required to have covers when not in use."[7]

Some Texas cities limited water use during droughts in 2023. Katy, Texas,

Skip the Disposal

Running the garbage disposal uses 2 to 5 gallons (8–19 L) of water per use. Experts advise people to avoid using this appliance. Instead, they should scrape food into the trash. Fruit and vegetable scraps can be composted.

residents could only water their lawns at night. They were not allowed to add any new plants to their yards during the drought. Rosenberg, Texas, closed its public splash pad.

SAFETY

People need to take special steps to stay safe during droughts. Cities in drought areas often announce burn bans. In Dallas, this means that people cannot start open fires. Burning brush is not allowed unless the brush is in an enclosed container. No sparks can be allowed to escape.

Sometimes fireworks displays are canceled during droughts. This might mean people cannot use backyard fireworks. But citywide displays could be canceled

Around 19,500 reported fires are caused by fireworks in the United States every year.

as well. In 2023, parts of Missouri suffered from a drought. Tim Bean was the Missouri State Fire Marshal. He warned people against using fireworks. He said, "We are alarmed at the conditions, and if people aren't wise . . . we could see some dramatic situations."[8]

Personal safety includes being aware of the heat that often accompanies droughts.

People need to have a place to cool off
when needed. They should be sure to
eat healthy foods. They should drink
enough water.

People also need to know whether
their water is safe. When there is less rain,
things that **contaminate** water become

Boiling water kills germs, but it won't necessarily make water contaminated with toxic chemicals safe to use.

Using water thoughtfully is important whether or not drought conditions are present.

more concentrated. Local governments may tell people to boil their water. Water should be boiled for at least 1 minute. This will kill microorganisms that can make people sick.

LEARNING ABOUT DROUGHTS

Unlike other natural disasters, droughts are not sudden. They are slow to form. But they can be deadly.

Fortunately, the threat of droughts can be reduced. People can take steps to fight climate change. They can learn about water conservation. When a drought strikes, people can work together to conserve water until the drought breaks.

GLOSSARY

atmosphere
the layer of gases that surround Earth

conserve
to protect a resource from waste and overuse

contaminate
to make something impure or polluted

dormant
temporarily inactive

greenhouse gases
gases that trap heat in Earth's atmosphere

infestations
presences of large numbers of insects or other pests

insecurity
uncertainty

microorganisms
tiny organisms invisible to the naked eye

particles
very small pieces of something

precipitation
water that falls from the sky

SOURCE NOTES

INTRODUCTION: THE DUST STORM

1. Quoted in John O'Connor, "At Least Six Dead in Dust Storm Crashes on Illinois Highway, Police Say," *Washington Times*, May 1, 2023. www.washingtontimes.com.

CHAPTER ONE: WHAT ARE DROUGHTS?

2. Quoted in The Weather Channel, "Science Behind Drought," *YouTube*, June 27, 2013. www.youtube.com.

CHAPTER TWO: THE EFFECTS OF DROUGHTS

3. Quoted in KCAL News, "Drought, Extreme Weather Heavily Affecting Farming Industry," *YouTube*, September 30, 2022. www.youtube.com.

4. Quoted in Jessica Merzdorf, "A Drier Future Sets the Stage for More Wildfires," *NASA*, July 9, 2019. https://science.nasa.gov.

5. Quoted in PBS NewsHour, "Utah's Great Salt Lake Shrinks to Unsustainable Levels amid a Decades-Long Megadrought," *YouTube*, October 18, 2022. www.youtube.com.

CHAPTER THREE: DEFENDING AGAINST DROUGHTS

6. Quoted in Clare Amari, "Drought Conditions around Houston Worsen, Prompting Water Restrictions and Burn Bans," *Houston Landing*, July 29, 2023. www.houstonlanding.org.

7. Quoted in SaveWaterSB, "City Declares Stage 2 Drought," *YouTube*, May 27, 2014. www.youtube.com.

8. Quoted in FOX 2 St. Louis, "State Fire Marshal Urges Missourians Not to Set off Fireworks amid Drought," *YouTube*, June 29, 2023. www.youtube.com.

FOR FURTHER RESEARCH

BOOKS

Tammy Gagne, *Wildfires*. San Diego, CA: BrightPoint Press, 2025.

Kathleen A. Klatte, *Droughts*. New York: Rosen, 2023.

Steve Tomecek, *All About Heat Waves and Droughts*. New York: Children's Press, 2021.

INTERNET SOURCES

"Drought," *Ready*, March 21, 2024. www.ready.gov.

"Drought," *World Health Organization*, n.d. www.who.int.

Alison Ince Rodgers, "Understanding Droughts," *National Geographic*, April 3, 2024. www.education.national geographic.org.

WEBSITES

Drought.gov
www.drought.gov

Drought.gov is a central point for information on droughts in the United States. People can find information on droughts by region.

National Weather Service
www.weather.gov

The National Weather Service is an official US government site that provides information about weather, including droughts.

US Drought Monitor
www.droughtmonitor.unl.edu

The US Drought Monitor provides detailed maps of droughts in the United States. People can use the site to see if there is a drought in their area.

IMAGE CREDITS

Cover: © Piyaset/Shutterstock Images

5: © Luiz Ferreira/Shutterstock Images

7: © Mohammed_Al_Ali/Shutterstock Images

8: © Ververidis Vasilis/Shutterstock Images

10: © Kent Weakley/Shutterstock Images

13: © Kunal Mehta/Shutterstock Images

14: © TM Creations/Shutterstock Images

17: © VectorMine/Shutterstock Images

19: © Rich Carey/Shutterstock Images

21: © kavram/Shutterstock Images

22: © Everett Collection/Shutterstock Images

25: © yelantsevv/Shutterstock Images

26: © tawanroong/Shutterstock Images

29: © You Touch Pix of EuToch/Shutterstock Images

31: © Real Window Creative/Shutterstock Images

32: © altanakin/Shutterstock Images

35: © Andrew Zarivny/Shutterstock Images

36: © Paolo Bona/Shutterstock Images

39: © Randy Judkins/Shutterstock Images

40: © ARSimonds/Shutterstock Images

43: © MD_Photography/Shutterstock Images

45: © Vach Cameraman/Shutterstock Images

46: © igorwheeler/Shutterstock Images

49: © Dean Drobot/Shutterstock Images

50: © Peppersmint/Shutterstock Images

54: © Alik Mulikov/Shutterstock Images

55: © New Africa/Shutterstock Images

56: © Mark Fisher/Shutterstock Images

ABOUT THE AUTHOR

Sue Bradford Edwards is a nonfiction author who lives in Missouri. She is the author of several Brightpoint Press titles, including *Become a Construction Equipment Operator*, *What Are Learning Disorders?*, and *Robotics in Healthcare*. Her father grew up in the West Texas desert. He taught her the importance of water conservation. She and her family set up rain barrels to collect rain to water a community garden.